Absolute String Theory

An Examination of the Fundamental Units of Nature and How They Interact to Affect Reality

by

Miroslav Halza

Contents

Prologue
Universal Speed Limit
Strong Nuclear Force
Mass
Gravitation
Electromagnetism
Elementary Particles
Bosons
Conclusions

Prologue

In this monograph, I argue that String Theory should use *all* the properties of strings in its conception, since it purports to explain subatomic phenomena as the results of the interaction of elementary strings and the structures they comprise. All the geometric shapes and physical properties inherent to strings should be included in its theoretic framework. These characteristics should contribute to determining the structure of subatomic particles and manifest in all forces in nature.

Strings are essentially one-dimensional objects in space, and my view is they are the live "units" of nature, the fundamental structures underlying all the rest. The density of string points on a line can change, and so they may elongate and shorten. Furthermore, they can form not just straight lines, but also curves. Strings may have open ends or may be closed into loops. Their intrinsic properties include energy density and tension. We register them macro-worldly in music, when they form many variants of standing waves created by their vibration, spin, or both.

When elementary strings propagate through space, they do so in the form of waves. There are two basic forms of traveling waves. Strings may form **longitudinal** waves, vibrating in the direction of travel, or **transverse** waves, rotating or vibrating perpendicular to the direction of travel.

Having the property of speed gives these traveling strings another property, generally known as momentum or kinetic energy. In accordance with physical law, this property can be changed during their interactions with other strings that they encounter as they propagate, which is why their collisions create forces that can repel or attract objects toward certain directions. In addition, strings may bring their intrinsic energy to physical objects when those objects absorb them or otherwise encounter them, if the collisions are imperfectly elastic. Forces carried by longitudinal and transverse waves effect even distant objects, because they can propagate even through a vacuum, and therefore are clearly the carriers of the universal fundamental forces.

Because strings also exert forces on nearby objects, they represent local forces that are very familiar to us as well. The shapes of the strings involved play a critical role in this situation. If some string shapes are in dynamic states, they can influence other strings either by attracting or repulsing them. If these shapes lie at the surfaces of subatomic particles, then such particles exert their total force there. That's why we have three kinds of primary particles in relation to local forces. The first is neutral, the second positive, and the third negative. The influence of the strings they carry form a field of that particular force.

Let's take a closer look, starting with String Theory as proposed by modern-day theoretical physicists.

The Universal Speed Limit

String Theory arose in the period of 1968-70 as an attempt to understand the strong nuclear force in the Standard Model of subatomic particles. At that point, the physics of strings began to play a primary role in our understanding the nature. However, the physicists developing the model implanted some limits in it and, later, also included some non-physical additions that weakened the Standard Model. The first limit is the requirement that the speed of light in a vacuum be the maximum speed allowed in nature (per Einstein). This means that the fastest speed that any propagating object can reach is the speed of light, **c**, which is 186,000 miles (299,792 km) per second.

However, the speed of light is merely the speed limit *for objects that human beings can detect*. Our ability to discover how nature works is filtered through products of nature (flesh); and therefore, we get "manufactured" results, not the fully realized results as experienced by the universe itself. Thus, our perception of nature is unequal to the *reality* of nature.

Einstein's speed limit does not apply to how a string propagates—for example, via transverse waves in a vacuum. An equilibrium point of vibration, or a center of rotation, travels one-dimensionally at the speed of light, **c**. However, if a string vibrates perpendicularly to its propagation direction, the resulting trajectory is two-dimensional rather than one-dimensional. As the equilibrium point travels in the straight line, the vibrating point creates a sinusoidal wave. Therefore, the velocity of a vibrating point should be added to the velocity of the equilibrium point, which results in a velocity *higher* than **c**. This is why the resulting velocity of a string must be higher than **c**, since the sinusoidal trajectory of any point in a wave is *longer* than the straight-line trajectory of the equilibrium point. The resulting velocity of string points traveling by transverse wave is thus higher than Einstein's speed limit.

Similarly, when a point rotates around an axis, it has a rotational velocity. If the center of rotation travels at the speed **c,** then it creates a line (axis) around which the point rotates. This is the case of light. Clearly, points along such a string must have a velocity higher than the traditional speed of light.

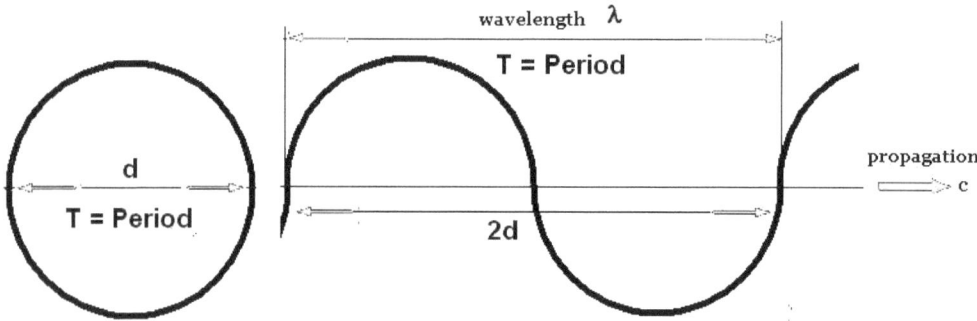

The speed of spherical sinusoidal waves is equal to the length of an axis during one rotation around the axis, and so to a wavelength, **l**, created during time, **T**, needed for one cycle. Hence, the speed of our wave is equal to a wavelength divided by a period of one rotation, $v = l/T$. Yet **l** is equal to **2d** (d = diameter of a circle), and therefore $v = 2d/T$ also. Light propagates by spherical sinusoidal waves, and so $v = c$ in a vacuum.

However, the rotation point changes its position not just in relation to the propagating center, but also to its rotation around the center. The length of one rotation around the center is equal to the length of the circumference, which is equal to **pd**, where **p** is pi, a number approximately equal to 3.141529. During the interval of time, **T**, the orbiting point has the speed **pd/T** around the centerpoint. The resulting velocity of a rotating point is a sum of both velocities, and therefore must be over **p**/2 of the speed of light (or about 1.86 **c**). Therefore, the speed of points in strings in the subatomic world should be over 1.57 times higher than the normal speed of light in a vacuum.

Similarly, as a string changes the density of its points, it vibrates in the direction of propagation as a longitudinal wave. The lengthening and shortening runs in a frame, which travels at the speed of light, **c**. That is why the interior vibration should contribute to **c;** and so we get the real maximal speed of the interior of the longitudinal waves. This assumes that a point of a string (its segment) may also propagate one-dimensionally at a speed higher than the speed of light in a vacuum.

Therefore, it's obvious that Einstein's speed limit must not be valid in nature. If it were, our world would not exist.

Why not? Early theoretical physicists knew of only one force that affects all celestial bodies, and concluded that the Universe would not have existed without this force. They believed that the universe as we know it came about due to the effects of gravitation. Indeed, some physicists view gravity as omnipotent and omnipresent (Stephen Hawking considers gravitation to be a kind of "god" of the universe). Thus, they considered gravitational attraction a force controlling the universe. Still, they were aware that a force can exist *only* as a result of some interaction between objects. Observing moving celestial bodies forced them to accept that there exists some invisible object interacting between the bodies visible in the universe. The particle that carries this force they named the **graviton**.

This force carrier has to propagate from one body toward which another body is pulled. This string propagates without any interruption, because the universe is mostly empty space. Therefore, the speed of light, **c,** applies to gravitons also. If gravitons travel at the speed **c**, then gravitational force propagates at **c** as well.

Because gravitation means attraction, the corresponding object is pulled toward the source of the gravitons. Thus it should be obvious that gravitational force results from an interaction between a celestial body and the elementary strings (gravitons) causing that gravitation. According to Newton's Third Law, when an interaction between two objects causes an attractive pull on one, there must also be a corresponding attractive pull upon the other. In other words, a celestial body attracted by a graviton attracts the string(s) comprising the graviton as well. If this attracted string propagates at **c** before the interaction, then during the interaction it has the speed **c** *plus the speed developed by the interacting force of attraction.* For this reason, if the gravitational force exists, then Einstein's speed limit must not.

Nonetheless, the speed of light *is* a limiting factor for human beings, since we were created in light (Genesis 1:3). We must not and cannot exceed the speed of light and survive. Consequently, time does not exist for us at speeds higher than **c**; i.e., time ends when we achieve light speed.

I understand that our time relates to our inability to keep up with light, and for that reason, we lose the exact picture of what's really happening in nature. This is because we're unable to cope *immediately* with the information brought to us by transverse waves, due to the processing limitations of our brains. Again, we require some time to handle the data, and so our responses are delayed. This phenomenon applies to photons, because processing photons causes a delay (the effective speed of light is always slower during its propagation through a material medium).

Consequently, time runs more slowly when we're moving very rapidly, and it seems to others that we have lived longer. On the other hand, if our processing ability were quicker than **c** as it relates to gravitation (i.e., collisions of gravitons with ordinary matter), then our time does not exist as such. If ordinary matter accelerates gravitons, then the sensation of gravitons is sped up. Hence, besides what we perceive as *our* time, there must also exist some other type of time that does not depend on our perception of it. As the speed required for processing gravitons is different from the speed required for processing photons, our time in the "world of photons" must differ from (their) time in the "world of gravitons."

Since graviton processing is faster than the reality we know, then there obviously exists another spacetime contiguous with our own. Hence, mathematical models of nature that use only our spacetime are invalid; mathematics won't work properly with any relative item, and our spacetime is relative. Calculations using our spacetime are useful only for allowing us to imagine how we sense nature.

In this context, "nature" means objects in the three-dimensional space we can directly perceive. If an object has a speed, then it just changes its position in our three-dimensional space. Changes in position in our space cannot change our space, in the sense of curving it geometrically. We need to measure and compare changes of position in three-dimensional space based on a standard. The standard is light. Still, light moves only in our three-dimensional space, and is useful only for comparing changes in that space.

Our main unit of time is based on the standard of the speed of light. One second is 299,792,458 light meters. If you change your position at a rate of 1 m per second, and therefore have the speed 1m/s, then you are delaying 299,792,457 light meters. But if you move 299,792,458 m per second, then your delay in relation to light does not exist—and so (our) time does not exist for you.

Hence, moving in three-dimensional space cannot create the fourth dimension (we can use a fourth dimension only for the supportive measuring of movements, i.e., of changes in three-dimensional world). The conclusion here is that nature runs independently of human observation. The speed of light is useful only for relative measurements, not for creating accurate models of nature. Therefore time is just a relative dimension, and cannot effect the true dimensions of nature and cosmic reality.

The Strong Nuclear Force

In order to be valid, String Theory must also contribute to our understanding of the strong nuclear force. However, because the proponents of String Theory were initially unable to describe the strong nuclear force using their theory, they accepted another, unrelated theory instead. According to it, bound particles called quarks form protons and neutrons. Strong interactions exist among these quarks due to the combinations of three "colors" (as these properties were called), which led to the theory known as quantum chromodynamics. They then applied this strong interaction to the atomic nucleus. But because this theory does not consider any strings there, they almost abandoned String Theory. This begs the question: is quantum chromodynamics an accurate scientific theory? Is it a scientific theory at all?

Fundamental to our understanding of physical science is the Law of Conservation of Mass and Energy. As we understand it, energy cannot be truly destroyed; it can only change its state. Science has excluded everyone whose theories violate this fundamental law of physics. For example, we have rejected alchemists and those who have claimed to create perpetual motion machines. That is why when scientists divide matter into pieces, those pieces *must* add up to the total of the original mass and energy. This law applies to all inputs and outputs in physical changes. Therefore, when a theory states that a baryon (e.g., a proton or a neutron) can be divided into three parts, then all three products together must contain the same mass and energy as the original particle. They need not each weigh the same, but they *must* weigh the same as the original baryon when the weights are totaled.

The physicists who conceived the theory of quarks were aware of this requirement, and therefore assigned their subatomic particles a mass equals to one-third of the mass of a baryon—at first. Later experiments have disproved this assumption of mass; and that is why quantum chromodynamics, if it exists at all, *cannot* be built on quarks. For example, we know that a proton has a mass 938 MeV/c^2, but its three quarks have a total mass of about 11 MeV/c^2. This *strongly* violates the law of conservation of mass and energy, placing quantum chromodynamics at the same level as medieval alchemy.

To explain the strong nuclear force in terms of their theory, the proponents of quantum chromodynamics introduced a boson that causes this strong nuclear force. They call it a gluon, and claim that gluons are the exchange particles for the color forces between quarks. They claim the gluon's mass is essentially zero (or an experimental limit of 0.0002 eV/c^2 at most).

Some try to defend this violation of the Law of Conservation of Mass and Energy by saying the rest mass of the proton (98.5% of the total, less the measured mass of the "quarks") lies in the strong nuclear force. However, this itself violates the definition of a force. A force is a push or pull on an object resulting from the object's interaction with another object. Therefore, a force among objects or particles never rests; it always acts on the object or a particle. Needless to say, such a force *cannot* contribute to the missing mass. Apparently, quantum chromodynamics is little more than a philosophy of physicists wearing colored glasses, and therefore has no place in the real world.

If you call something "String Theory," then the *whole* theory should work via the mechanism of strings, and not be camouflaged with point particles like quarks. As already discussed, elementary strings should propagate as waves at the speed **c**—but theoretically, there should also exist single strings that are *not* propagating, i.e., that are at rest. Consequently, there should exist strings bound into clusters or bunches, thereby creating material objects. So quarks, if they exist, must be assemblies of strings—not the proposed point particles.

To defend their theory, the quantum chromodynamicists appeal to the apparent detection of quarks in colliders when baryons are bombarded by electrons. It's unlikely that a baryon can be cracked into neat individual pieces by any such bombardment; rather, it should be broken into bunches of strings and large numbers of individual strings. As an analogy, no one would suggest that a random bombing would divide a house in three equal parts, since a result of any bombing are dust (small particles), pieces of bricks and wood, individual bricks and planks, and larger, intact conglomerations of building material. Whole portions of a house might even remain intact—for instance, walls or cellars. The same should occur when breaking apart baryons.

Thus, they had to realize that some of the smallest particles (for instance, photons) escaped, while clumps of strings also remained in the debris of a baryon. These blocks, which they define as quarks, are just sets of many strings, and therefore never have the same mass. Repeated collisions of electrons with baryons do not produce the same types of quarks. They have in fact confirmed that identical quarks do not exist, since there is a range of mass for each quark. For the example, the "up" quark has a mass 1.7 to 3.3 MeV/c^2, while the "down" quark ranges from 4.1 to 5.8 MeV/c^2. Their lifetimes do not exceed the time it takes light to travel through the diameter of a baryon; thus their lifetimes are about 10^{-15} light meters.

Since these "quarks" exist only in the volume of a baryon and cannot exist outside it, they cannot actually be individual particles. True particles must exist in "naked" form, as identical single points with the same mass. The real impact of these experiments is that the researchers conducting them detected many *fragments* of shattered baryons, selected the biggest ones, and named them quarks.

Further argument for this interpretation comes from the fact that the quark theorists assign odd electrical charges to their erstwhile particles. The up quark has a proposed charge of +2/3, the down quark -1/3 (the unit charge is the charge of the proton). However, they have never actually *observed* such fractional charges, which is why I've concluded that their theory is incorrect. Consider this: quarks come from the same baryonic environs, and therefore should not bear dilemmatic charges—and should not mix positive and negative charges.

Would our philosophers abandon quarks, or at least the theoretical charges for quarks, if science were to find reality to be different that what they've convinced themselves it is? Probably not, because in their minds, the space in human knowledge for information about the charges of quarks is already occupied. Therefore, their propaganda has already subdued our world to their misinterpretations, and the truth will not change their conclusions.

Other physicists have found that a neutron can be composed of a proton, an electron, and unassociated strings (photons). As they studied neutron decay, they attempted to balance an equation: neutron = proton + electron + photon, according to the conservation of mass and energy. But they could not balance it, and therefore concluded that the rest of the mass must be carried on an undetected particle. The final result of real physics is that the neutron decays to form a proton + an electron + an (anti) neutrino + a photon (a string propagating at **c**).

As small as they are, the proton and electron are huge in relation to any string or strings bearing energy quanta (e.g., photons). Electrons can lose photons, as seen in the emission spectrums of chemical elements, so there are *loose strings* in the subatomic environment. For that reason, the elementary particles and quasi-particles (which is what quarks are) must be composites of strings. These composites of strings comprise ordinary matter. Strings are bound into the volumes assigned for them according to their dynamic properties, and thus give rise to subatomic particles like protons, neutrons, electrons, and neutrinos. When these elementary particles are bound together, they create atoms, the building blocks of matter.

The density of strings in matter varies, so three basic kinds of matter exist. Since the universe is huge, let's look there for all the possibilities.

Ordinary matter, as we know it on our Earth, is characterized by electrons orbiting protons and neutrons in the nuclei of atoms. The smallest possible atom consists of one electron orbiting one proton—a hydrogen atom. This is by far the most common type of atom in the universe. The maximal density of this kind of matter can be reached in a white dwarf, where its state is dependent on boundary electron degeneracy pressure. Protons and neutrons can even form plasmas there, covered by electrons. Thus, original matter has electrons on its surface.

The second kind of matter is that which is condensed into atomic nuclei. Free electrons do not exist here, because they have been suppressed into nuclei; the matter has lost most of its volume, i.e. the free space where electrons once orbited atomic nuclei.

Therefore, this matter has a much higher density than ordinary matter. If the diameter of an atomic nucleus is 10^{-15} meters, and the diameter of an atom is 10^{-11} m, then the density of nuclear matter can be up to 10^{12} times higher than the density of ordinary matter. We see this kind of matter in neutron stars, which have a density equal to the density of atomic nuclei (3×10^{17} kg/m^3). Hence, the second kind of matter consists only of neutrons.

Lastly, astrophysicists theorize that a black hole forms when the remains of a dead star collapse under its own gravity, squeezing even subatomic particles together. Therefore, black holes are the densest objects in nature. That's why the most compressed matter created by supercolliders (like the Large Hadron Collider, or LHC) should also reach the same density as matter in a black hole. The "Higgs boson" is the tiniest particle in the volume now observed, but theoretical physicists are now predicting the existence of even smaller particles that make up the Higgs ("techni-quarks"). In any case, the smallest particle of this kind of matter is a collapsed neutron.

The collapsed neutron may very well be the densest form of matter in existence. Strings bound into the volume of a neutron still have enough room for movement, but strings in a collapsed neutron are packed so tightly they cannot move, creating the densest concentration of strings in the nature.

A simple string can oscillate in two ways. We can see the first type of vibration at the ends of a string that is not fixed. A distance between them (i.e., the distance between nodes) increases and decreases, and so balances between the maximal and the minimal length of a straight-line string as an oscillating object, as with a spring. The second kind of oscillation occurs in the transverse direction, where the maximal displacement is in the antinode. We see something similar at a membrane. Certainly, we should suppose that each vibrating string has an impact on neighbor strings, especially inside a collapsed neutron, because the strings are packed into a very small volume.

The first kind of vibration of the first string influences a neighboring string (spring) and they may couple and so on, creating a longitudinal vibrating thread. Half of the wavelength represents one string (spring), and so our thread is a combination of these strings, with some invisible bond between them. Such a set of bound strings (a thread) does not create stability for the strings at the end of a thread, and therefore they may be pulled loose if they lie on the surface of a particle. However, pulsing antinodes in the transverse direction are pulsing into other threads, and this holds strings inside an object (in other words, the strings cling to one another).

Forces acting on the strings inside an object don't act alike on strings at the ends of threads. An attractive force arising from the pulses acts on them from the inside. This gives rise to surface tension, creating the same result as the surface tension of a drop of water. There, molecules are pulled equally in all directions, forming a spherical drop. That is why particles of matter are spherical particles.

The logical conclusion is that strings are bound into matter due to their ability to vibrate into two dimensions or even into three, and so interpenetrate with neighboring strings. In forming a nucleus, protons and neutrons come so close one another that the surface strings of one interpenetrate into the surface strings of the other, "gluing" them together. This explains what we call the strong nuclear force. This force results in the compatibility of strings with the volume of a baryon, and the compatibility of the baryons themselves (i.e., protons and neutrons) to some volume of a nucleus in an atom.

Mass

Particles like protons, neutrons and electrons all have specific masses. But individual strings traveling in space lack mass. Given this fact, it should be clear that the property we call mass arises due to relationships among strings inside a volume of matter.

As previously discussed, the densest known form of strings occurs in a collapsed neutron. And yet, I don't believe that these strings are so compressed that they're entirely unable to vibrate. I think that, instead, they are simply woven into that matter, as fibers are into a fabric. Hence, the matter is in a material form much like fibers woven into a rug, and the strings remain "live."

Ordinary sets of strings gather in particles like neutrons, protons and electrons. Strings can move within those particles, and I would compare their state with that of a gas—for example, propane or butane. The molecules of a gas possess the kinetic energy of motion, which is comparable to Brownian movement in a gas. This zigzag motion may occur only in one dimension, in which case we would have motion as seen in a spring or pendulum. In this case, the particle swings between two maximal displacements from its equilibrium point.

If a string is swinging this way, then by necessity it possesses the property of motion. But where does it get the energy for motion? First, when it was displaced sideways from its resting equilibrium position, a force acted upon it for a while, pushing it; and so the string received its momentum. But I suggest here that the string moves because it brought this initial momentum into the fabric of the particle when it joined the particle. Our string had a momentum of the speed c when it impacted the particle. This initial momentum is the reason strings vibrate in matter, and it is the potential for "traffic" inside protons, neutrons and electrons, and thus is the potential energy for back-and-forth movement. The maximum potential energy occurs at the point of the maximal displacement, just as the direction is changing. The speed drops to zero at maximal displacement, and so the momentum is zero too. Zero momentum could occur only after a head-on collision with another particle or string of the same momentum, so that the momenta are canceled.

In the case of spring-like motion, the back-and-forth movement is not the result of a collision. That is caused by the potential energy of the spring (or, for a pendulum, its gravitational potential). Since momentum is conserved and cannot disappear, then at the maximal displacement, there must be some potential to initiate new movement back toward the equilibrium point. Thus, strings possess energy for motion inside elementary particles of matter.

Besides, when a transverse vibrating string (a photon) strikes an electron, the photon brings a quantum of energy into the electron. Photons are waves having both frequency, f, and wavelength, l, and their energy is equal to fl. This means that a string that vibrates in an open space vibrates the same way in a closed space, such as the volume of an electron. Nor does this energy change in a collapsed neutron; the quanta of energy remain.

Hence, the string possesses both a quantum of energy and the energy of motion (vibrant motion), as previously outlined. That is why we need to distinguish between these energies.

Vibratory energy of motion switches back and forth from potential to kinetic energy. However, the initial reason for this energy is the inherent property of strings to possess momentum—to travel at **c**, the speed of light. Some theorists (such as Edwin F. Taylor) unify momentum with energy into the physical property of "momentum-energy" in relation to spacetime. I believe the term "momentum-energy" is in fact a good explanation for the vibratory motion of strings. Hence, particles of ordinary matter (i.e. protons, neutrons, electrons) consist of many strings, all moving inside the volume of the particle. The collective movement of the strings creates the internal property of baryons. This physical property in relation to forces, velocities, and momenta is vectorial; and therefore, when there is no interaction with the outside world, they all cancel each other out.

Normally, energy is not vectorial. It is a quantitative property, and that is why we can determine the sum of the momentum-energy in a particle—and why physics must reckon with it. Physicists must name it, explain it, and calculate its quantity. Indeed, they have named it mass, and calculate its quantity in units like pounds and kilograms. So they have already explained this property, but only in a metaphysical sense—and metaphysical explanations are not accepted in real science.

They see this physical property, in other words, as interacting with our material world as some thought it did during the Dark Ages. They have introduced a god particle into their theories: a mysterious particle that lurks outside matter to solve this hidden property of matter. History has already proven similar pseudophysicists wrong when they explained heat as the results of "caloric," another mistaken particle invented to explain an inherent dynamic property of particles.

The proponents of the Standard Model added mass to Spring Theory by inventing what they call the mass field (the Higgs field). Since a field must consist of particles, they therefore suggested that this field arose from an already-named theoretical particle called the Higgs boson. When a particle with the extreme density they required was found, they adopted it for their mass field, to legitimize their mathematical philosophy. They celebrated their final understanding of nature by using the LHC to transform ordinary matter (protons) into extraordinary matter to fill the Higgs/mass field. Who gave them the right to twist the results of these experiments in such a way? Is their logic even a human logic?

They claim to have found the particle that gives mass to protons. Yet the truth is, this particle was not simply discovered, but *created from protons* smashed together in the collider. As with all crashes, smashing protons together just demolishes the assembled material. Therefore they created a crash product, a wreck, which consists of strongly packed matter. As a wreck, it contains the same components that were in the objects *before* the crash, as components of protons cannot be changed into other components during or after the crash. If strings comprise the substructure of protons, then the crash product also contains strings; they're just assembled differently after the crash, more tightly packed. Therefore, it's nonsense to claim that a proton crash product—the so-called Higgs boson—affects intact protons in such as way as to give them mass. Simply put, pieces of smashed protons cannot generate the mass of other protons.

Why the deception? Because they needed to save the Standard Model, which in addition to failing to properly explain mass, fails to adequately explain any force among universal bodies, and does not explain gravitation, "dark matter" and "dark energy." Rolf

Heuer (CERN's Director-General) and his colleagues have credited themselves with the ability to change ordinary matter into astonishing, extraordinary matter—into the medium of an invisible world. They claim that their theory that proton crash products generate the mass of protons is a confirmed fact. They claim, too, that this new god particle, the Higgs Boson, confirms the String Theory and the Standard Model as valid—and made a celebration of it at the Nobel Prize Award Ceremony.

The theory underlying the Standard Model wants to be a string theory—and therefore, mass should be understood as an intrinsic property of strings. If a string has energy, then two together have an energy that is a sum of both strings, and a whole collection of them possess a sum of all the string energies. If a string has momentum-energy, then two bound together possess the sum of both momentum-energies, and a whole volume full of them has a sum of all the momentum-energies of all the strings within. In fact, a particle created from strings bears the total energy (quanta) of all the strings built into its volume, as well as the total momentum-energy of their movements inside that volume.

Therefore, momentum-energy is the total kinetic and potential energy of all the quanta of energy within a given volume of ordinary matter. This momentum-energy is expressed as that volume's mass.

Gravitation

Earlier, I pointed out that the velocities, forces, and momenta of strings are vectorial inside matter, and that therefore, they cancel each other out. However, if we bring these vectorial quantities from the outside world to them, then these external quantities interfere with the internal equilibrium—and we see that matter (as a whole system) has a force, a velocity, and a momentum. Hence, ordinary matter may interfere with external forces like velocities and momenta because its own particles also exhibit forces, velocities, and momenta. That's why nature doesn't need any additional tools to break the symmetry of protons, electrons, and neutrons to make them capable of interfering with the Higgs field and so acquiring mass. Specifically, according to mathematical philosophy, particles lack mass because they are in symmetry.

Free strings must have a momentum in order to enter material objects, which is why matter interacts primarily with strings moving at light speed. Again, the momentum-energy (mass) inside matter responds to the moving strings, and thus to the field possessing gravitons. Hence, the mass responds to a gravitational field. We can measure the resulting response by weighing an object. If the intensity (density) of the moving strings is constant, then by doing so we register the proportional force related to the interior momentum-energy (mass) of the weighed object. If a force acts on an object and causes an external movement, then it is an accelerating movement ($F = ma$). That is why the gravitational constant has in the past also been referred to as gravitational *acceleration*. This gravitational acceleration differs in reference to one's location in the universe (e.g., it's different on the Moon than it is on the Earth). Furthermore, gravitational acceleration decreases as we move farther away from a particular object acting upon us. This suggests that the moving strings creating the gravitational field of the Earth emerge from the matter comprising the Earth. Therefore, our gravitational field is inherent to our world, since it arises from ordinary matter and influences other ordinary matter.

But according to Einsteinian physics, gravity is a result of curved spacetime—or more precisely, the geometry of spacetime is a manifestation of gravity. In spite of their understanding of gravity and spacetime curvature, the primary mistake of the Standard Model theorists is in creating a need for the expansion of spacetime in order to explain the expansion of the universe. If our space is expanding, then what is it expanding into? It must be expanding into some other space, and they into a third space, and so on *ad infinitum*. Or, if our spacetime is expanding into some other spacetime, and they into a third spacetime, and so on, then other times must exist there.

String theorists resurrected String Theory with their theory of quantum gravity. Still, it appears that they have merely resurrected the carrier of the gravitational force by positing a bosonic string called a graviton. This leads to Bosonic String Theory, which requires 26 dimensions in its mathematical model (twenty-five are spatial, and one is temporal). Needless to say, this is contrary to human experience—that is, it does not fit into our real world. Our world typically classifies any theory of another, invisible world as religious; and so those who work with 26 dimensions are glorified as giants in science.

What giants know, we do not know. But ordinary men and women know that a gravitational field exists in the world in which they live, and that's how they know their weight; they know the force acting on them. Still, this force is a physical property in our three-dimensional world, not in other worlds or dimensions. The attraction emerges from ordinary matter, in this case the Earth—and this is the real physics that applies to them.

The assumption is that strings emerging from the matter of the Earth affect our bodies without interacting in other dimensions, worlds, or universes. The density of these strings in a gravitational field cannot arise from any curvature of spacetime, because the speed of gravity is equal to the speed of light, and propagating light creates straight-line trajectories that allow us to see deep into the universe. We feel the Earth's gravity because the subatomic string structures emerging from the Earth travel through three-dimensional space and affect the strings making up the matter in our bodies.

The proper mathematical model for gravitation, then, is Newton's law of universal gravitation. This law demonstrates that the gravitational force between two objects is directly proportional to the masses of both objects involved. In other words, according to String Theory, the total number of outbound strings is directly proportional to the quantity of strings assembled in an object. The gravitational force is inversely proportional to both distances, and thus to the square of the distance between them. This proves that the strings producing gravitational force are outbound from an object. Their density falls in proportion to the distance from the object due to their spreading out into all three dimensions evenly. Hence, the true theory of gravity has its mathematical expression, and so has its laws.

The question is, what kinds of waves or strings create the gravitational force? We already know that a traveling transverse string is a photon. Photons do not interact with matter to create gravitational force, so they can't be the "graviton" we're looking for. Since we have just two possibilities for the propagating oscillation (transverse and longitudinal), then strings producing longitudinal waves must be those that create the gravitational force. In other words, a string causing the gravitational force must vibrate in the direction of its propagation, like a spring. Therefore, we may use the physics of springs to explain the gravitational force.

Theoretic physicists have invented point particles, quarks, to fill the volume of a baryon. This means that quarks vibrate inside protons and neutrons, typically in threes. But applying a very small string of the Planck length (1.6×10^{-35}m) to a quark having a diameter about 1.6×10^{-15}m is impossible to display, which is why we'll never be able to illustrate gravity using this assumption. However, if we implant a matrix of strings inside a baryon, it instantly changes the situation. Gravitation becomes the physical interaction of moving strings (elementary springs, if you will) with strings/springs inside a baryon.

To put it simply, inbound propagating strings collide with bound strings inside matter. Their collisions should be "sticky," resulting in no loss of energy. The spatial interpretation of this interaction can be imagined by hanging a string (graviton) onto a vibrating string, causing it to propagate side by side. We can derive other techniques from the dynamics of vibrating springs, as I showed earlier, creating longitudinal threads. In any case, strings are able to interact between themselves, and therefore the gravitational pull exists. Recall that when they are stuck together, they travel at the speed resulting from both momenta.

It is necessary for the strings we call gravitons to emerge from matter at a rate proportional to the number of strings inside an object. A physical law must be assigned to this emission. The Second Law of Thermodynamics determines spontaneous actions, and that is why this law should also determine spontaneous delivery of strings for the gravitational field of the object.

Vibrating objects are inherently dynamic. The First Law of Thermodynamics states that the total energy of an isolated system cannot be created or destroyed, per the Law of Conservation of Energy; it can be only be changed from one form to another. For vibration, there is the kinetic energy of vibration that is transformed into the potential energy of vibration and vice versa.

According to the Second Law of Thermodynamics, an isolated system evolves spontaneously toward thermodynamic equilibrium—the state of maximal entropy, or disorder, in the universe. This means that if dynamic particles are on the way to reaching their thermodynamic equilibrium states, they must release some of their energy as entropy. As far as swinging energy goes (as in a pendulum), the outgoing string bears the kinetic energy of swinging related to speed **c**. However, the string already has some potential energy arising from its displacement from the equilibrium point. This potential energy is lost, because it will not be used to force the string to return to the equilibrium point, representing the entropy of our spontaneous action. (Since energy cannot be lost, it accumulates in the universe, and so universal processes will run spontaneously until the maximum entropy is reached)

Matter, as I've described above, can exist in nature in two forms: as uncompressed ordinary matter and as highly compressed, collapsed neutronic matter. The universe confirms the existence of highly compressed matter in black holes. The strings inside a black hole have very little freedom of movement, because the volume of the matter they comprise is severely constrained. Thus, gravitons cannot apply their gravitational pull on matter in a black hole. However, due to the extreme restriction of their momentum-energy by pressure from above, there is a higher rate of graviton string production by a black hole than occurs with ordinary matter—similar to the way in which more molecules of water evaporate from warmer water, or more molecules of air leave a compressed tire.

Physically, the Second Law of Thermodynamics explains that more gravitons are released from matter the more it is compressed. Therefore, gravitons escape from a black hole faster than they do from ordinary matter. In other words, black holes expel the carriers or "messengers" of the gravitational force more readily than ordinary matter, which explains why black holes have a much stronger gravitational pull than ordinary matter of equal mass.

This means that the mass of a black hole cannot be established according to the law of the universal gravitation. Astrophysicists now believe that some black holes contain millions of solar masses, which should be wrong. My conclusions here prove the fact that enormous black holes are the centers of galaxies. The stronger attraction emanating from these black holes concentrates other celestial bodies into the systems we call galaxies. On the other hand, the weak interactions of black holes with the gravitational fields of other celestial bodies keeps the universe from collapsing.

The most obvious aspect of a black hole is that even light cannot escape; hence the name. As I understand it, this is because strings propagating at light speed (c) in space interact (are absorbed) by strings of matter having higher speeds than c.

Consider the formation of a neutron star, where the speed on its surface increases rapidly to a high fraction of the speed of light; the next decrease of volume should result in a higher speed on the surface of a black hole. The speed of the strings comprising its volume should therefore exceed **c**, even high above the black hole's surface. Presumably, they drop back to **c** at the event horizon of the black hole. Beyond the event horizon, however, the speed of these strings should remain higher than **c,** which is why they inevitably capture the strings bearing light quanta (photons). Since their speed never drops below **c**, light has no chance to escape from an associated string. In the case of gravitation, a string of matter has caught the graviton for a while, when its speed was higher than the speed of the graviton. When the speed of the particles interact, the system starts to drop toward to its maximal displacement energy and reaches **c** or less, whereupon an affiliated graviton is released and continues on its way at **c**. But photons caught by the strings and particles of a black hole do not have this option.

Another scientific phenomenon we have observed in high-gravity situations is time dilation. This has been proven by the use of very accurate atomic clocks, which in fact show different times when placed in different gravitational potentials. A different gravitational potential results in a different density of strings able to manifest the gravitational force. Atomic clocks measure the vibrational frequencies of elementary particles in specific atoms of ordinary matter, which means those particles are rapidly moving back and forth around a specific equilibrium point. However, according to my understanding of gravitational force, longitudinal strings interfere with other strings inside matter; thus they create interacting couplets moving at the speed that results from conservation of both particles' momenta.

This increases the speed for the graviton in coupled particles, but decreases the speed of the associated string in a clock. This lower speed of the string inside the clock particle should lower the particle's speed and thus prolong its period for achieving maximal displacement (especially when the maximal displacement is in a position to strike a neighboring particle). Prolonged traveling time will naturally decrease the frequency of the particle affected by the gravitational field. Therefore, an atomic clock measures time more slowly in a denser gravitational field.

Electromagnetism

When used to explain the Standard Model of nature, String Theory employs only one-dimensional strings (springs). This is why String Theory is sometimes called a "theory of springs." But while strings are one-dimensional, we shouldn't view them as just straight-line strings. Points of a string have no fixed place in relation to their neighbors; instead, they present themselves as something like a live worm. Consequently, from a spatial viewpoint, we need to employ three dimensions (or at least two dimensions) to display their shapes accurately. Besides this, they *are* live, so there's energy involved.

The fact that a string is "live" has great significance in determining its shape. It follows, then, that differently shaped strings should have different properties (as do different sounds in music).

Strings can be either open strings, which have two distinct and unconnected endpoints, or closed strings, in which the endpoints connect to form a loop. A fundamental wave of a closed string (the first harmonic) occurs at just half the string's wavelength, **l/2**. This occurs when a pulse displaced upward at the left end of the string travels through the string toward the right end. When creating a crest as it reaches the end of the string, it is hitting the boundary of the wave, and therefore the crest is reflected back (this is known as the boundary behavior of a pulse). According to the law of reflection, the angle of reflection is equal to the angle of incidence, which pushes the reflected crest downward. So, the reflected pulse returns to the other end of the string, where it reacts as it did the first time. Since the Law of Conservation of Energy must apply, the dynamic motion of displacement will be repeated indefinitely. Upon repeating, however, the angles will decrease (especially the angle of reflection) until they disappear. When they disappear, the reflection disappears; and then a creating crest and a reflecting crest (e.g., upward and downward crests) create a string that appears circular. However, it just *seems* circular; in reality it's a semi-circular string.

If such a "live half" exists, it is naturally attracted to a complementary string. This creates an attractive elemental force. This being the case, what kind of a pulse should the other string have in order for them to be paired? I think we need to start from the introduced pulse to determine this. As we've already seen, the pulse of the first string displaces upward at its left end and travels through the string toward the right end; therefore the pulse of the second string should displace *downward* at its left end before traveling through the string toward the right end. The second string thus picks up where the other left off. When both reach the end of the string (the node), they reflect or continue circling back to the start. Both strings may be bound due to differing wave-crests at the nodes.

An object forming circles and rotating upon itself (i.e., spinning, in the same way the Earth spins on its axis), influences other nearby rotating objects. Neighboring particles have moveable axes, which allows them to adjust themselves so as to be paired according to their different spins. When objects have different spin orientations, they are attracted to one another. We can observe this when chemical covalent bonds form. For example, an atom of hydrogen has only one electron. Its electron spins, attracting the electron of another hydrogen atom that spins in the opposite orientation, creating a covalent H_2 molecule. The force inherent to this bond is the result of the paired electrons.

The attraction of opposing spins also allows the existence of chemical elements with higher atomic numbers, which are created during nuclear fusion in stars. Essentially, the protons in a nucleus (forming a positive charge) must be accompanied by an equal number of electrons orbiting the nucleus. For example, heavy elements containing 100 protons in the nucleus must have also 100 electrons orbiting the nucleus. Since they are electronegative, like electric charges repel each other. Yet in spite of this repulsion, different spin orientations in the electrons allow atoms with high atomic numbers to exist, forming variant properties of ordinary matter.

This force must have its roots in the subatomic world, which is why we cannot see it in the "macroworld" in which we live. We need to hold an object and feel its push or pull to recognize the force acting on it. One "macro" object having a similar effect is a magnet. A permanent magnet is an object made from a material, such as iron, in which the spins of the electrons are separated. Iron includes moveable electrons, however, which can adjust themselves to the demands of surrounding the magnet.

When electrons of one kind of spin orientation displace to one side of an object and electrons of another spin orientation displace to the other side, then the object becomes magnetized. Some objects remain magnetized permanently, a result of special processing in a powerful magnetic field. This is why a bar magnet possesses a magnetic force. We deduce that the force is there because of the different spins of the electrons comprising it—so we may conclude that spinning strings also bear the magnetic force.

Consequently, both bound and moving strings can bear this force, because when objects orbit a center and the center starts to move, they form not circles but transverse waves. If we stop their propagation through space, then they form circles again. These types of waves are called electromagnetic waves, and they are very common.

The word "electromagnetic" is a compound word coined from "electricity" and "magnetism." Hence, we need to explain electricity in String Theory as well. Just as magnetism exists in electrons and also among strings, so the existence of electricity in electrons predicts that electrical charges exists in strings also. In my opinion, we need not look for some pointed charge of electricity and so depart from String Theory.

Instead, we need to look at shapes and their interactions, just as I demonstrated that magnetism is a force associated with the interacting spins of strings bound in loops or circles. Therefore, electricity should also have its roots in some sort of interaction between geometrically-shaped strings. Again, they should be paired, since both charges have the same value and differ only in being either positive or negative.

Let's take a look at the shapes of strings that best fit the electrical force—shapes not yet discussed. We already know that a straight-line shape produces the gravitational force, and that we get those straight lines by elongating and shortening strings. A circular or elliptical shape works for the magnetic force. In the macroworld, we get an elliptical shape by vibrating a string with both ends fixed (thus generating the first harmonic). What we haven't looked at yet is the type of shape formed when one end of a string is free and the other is fixed.

The unfastened end of a string is free to move up and down in two-dimensional space when disturbed, and does so. If a string is rotating along its length and has one end free, then it creates a funnel shape or vortex shape, with a circular cross-section. This should generate an attractive force within the vortex, as occurs in macro-scale vortexes like tornadoes and whirlpools. The conical mouth of one funnel might therefore drag in the narrow stem of another funnel, creating the paired couplets typical of the fundamental forces. That is why I conclude that the attraction of a conical mouth to a stem and vice versa forms the fundamental electric force in the world of strings, i.e., in the microworld.

A spinning vortex mouth correlates to a spinning circle, which we've already assigned to the magnetic force; and the narrow stem of a funnel corresponds to the straight-line strings engaged for the gravitational force. Therefore, we should also look for a funnel-shaped force in some strings, forming circles and straight lines. There, a circular string attracts a straight-line string. In electricity, the positive charge is attracted to the negative charge, and so the straight-line shape is dragged inside the circle. In the case of magnetism, a circled string or loop is dragged perpendicular to the electrical motion, because it should orient itself into a position parallel to the existing circle. Therefore, the directions in which the electric and magnetic forces act are perpendicular to each other in electromagnetism.

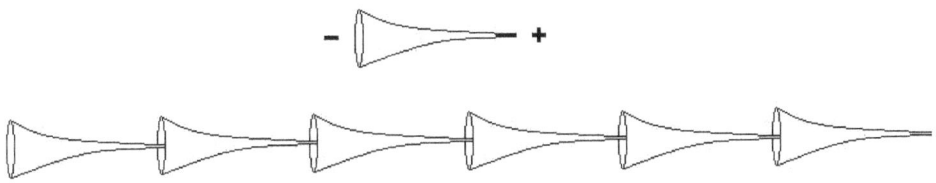

Elementary Particles

If elementary strings can create many different shapes, then their compositions must differ in order to allow them to do so. As a result, there should exist many internal structures in the particles of matter they form. The basic (elementary) structures, I believe, are electronic, protonic, and neutrinoless. Still, some bundles of strings are poorly arranged, and are therefore unstable, intermediate, and even create some rare particles (some exceptions always exist in the nature). A neutron should be categorized among unstable particles, since a free neutron decays with a half-life of about ten minutes.

So let's take a look at how arranging strings differently can create differences among particles of ordinary matter.

Gravitational strings really are strings, whether elongated and abbreviated. When they are bound, they are mostly bound inside neutrons and protons. Protons differ from neutrons primarily in that protons possess an electric charge; neutrons, as their name suggests, are electrically neutral. It may be that stems of funnel-shaped strings, as discussed in the previous section, are present on the surfaces of protons—those, or at least the straight-line ends of the component strings. If so, the stem end should contribute to the positive charge of the proton.

Consequently, the spinning, conical mouth of a funnel should represent a negative charge. These structures should be typical of the surfaces of electrons, which hold negative charges and orbit protons. However, inside the electrons there should be concentrations of straight-line strings, since electrons do possess both mass and a gravitational pull. The gravitational pull works for straight-line strings, but these being on the surface could be close to strings having one end fixed inside the electron, and another end loose on the surface. The loose ends may depart from the straight-line configuration as an electron spins. Thus, the surface ends of at least some strings should create the mouths of vortices as previously described.

Theoretical physicists postulate that there should be equal amounts of matter and antimatter in the universe. In other words, they claim, there are as many electrons with positive charges—a.k.a. positrons—as there are with the normal negative charge. But according to my string theory, particles having the volume of the electron are more likely to have strings ending with conical mouths at the surface (i.e., negative charges) than straight-line strings (with positive charges), and therefore, logically there cannot be equal numbers of positrons and electrons in the universe. Indeed, if they in fact exist, positrons are rare and unstable in any environment containing electrons. The lifetime of a positron is typically 10^{-10} sec. Similarly, proton-type particles would hardly have conical-shaped (negative) strings at their surface, particularly due to their size and lack of (enormous) spin, especially if they're bound by strong nuclear forces to other protons and neutrons in nuclei of atoms. Therefore, there could never be equality of numbers between protons and antiprotons.

If a neutron decays, then it should do so because it contains on its surface all kinds of string-shapes in some form of geometrical tension; that is, conical vortex "mouths" and narrow stems (straight lines) should occur side-by-side on the surface of the neutron. The best position for them is when a stem lies inside a funnel; and therefore this tension exists on the surface of the neutron. The stem *wants* to be in the conical mouth, in the same sense that a positive charge wants to be coupled with a negative charge, and vice versa.

As they try to remain so arranged on the surface of the neutron, some strings tear away from the body of the neutron. As they break away, they take with themselves other connected strings.

In this situation, we should presume that the loose collections of strings arrange themselves in morphological formations that achieve the best possible energetic state, and so form spheres. The result is the familiar stable ball-shaped particle. This particle has a very small mass (just 1/1839th of the neutron), and a negative charge equal to -1,602176565 x 10^{-19} coulombs. Strings, as they break from the neutron, have momentum, and so the new particle has a momentum; and that is how the electron acquires speed. The electron's attraction to the positive charge pulls it into a curved path around the proton as it otherwise attempts to move in a straight line; this is why the electron orbits the positively charged nucleus in an atom.

Meanwhile, other strings break away in ones or small groups, producing photons as the neutron decays.

Circular strings may also use the opportunity to break out of the neutron during neutron decay. These free loops form particles called neutrinos. Someone may point out that these loops *have* to depart along with the electrons, since the conical mouths of vortex-shaped strings do not differ much from circles. This is true, but the strings of electrons have conical mouths on their surfaces, while the other ends are fixed inside the particles themselves, among other things helping to create the mass of the electrons. Besides, they have different spin directions; i.e., the neutrino created during neutron decay has a different spin than the electron. For that reason, during beta decay an *antineutrino* emerges along with the electron created. The antineutrino differs from the neutrino by spin orientation only; so basically, it consists of looped or circled strings having a spin reverse that of the ordinary neutrino. There should be a few circled strings on the surface of the neutron, because the neutrino is a very small particle indeed. The created neutrino differs from the created photon in that the neutrino consists of bound strings, while the photon consists of a single string.

Certainly, looped or circled strings exist on the surface of the neutrinos. Inside, them, therefore, should exist some straight-lined strings, which vibrate and so give mass to neutrinos. However, they can be only very few of these, since neutrinos have an extremely small mass—less than one-millionth of the mass of the neutron, according to calculations. But I suggest that neutrinos may instead consist of looped strings only, and therefore do *not* have mass (as was originally theorized) and so do not interact with the gravitational field. However, neutrinos *do* interact with inductive matter very strongly due to their magnetism. That is why neutrinos may even surpass the speed of light, as was documented for Supernova 1987A. A burst of neutrinos was observed at three separate neutrino observatories approximately two to three hours before the visible light from SN 1987A reached the Earth.

After neutrino emission, the remnant of a neutron should be a particle with a surface cleared of irregular shapes; it would retain strings with simple shapes only—i.e., straight-line strings—and therefore resemble a hedgehog. We call this particle the proton, and it bears a positive charge equal to but opposite of the electron's. The ends of the strings on the surface of the proton should not change shape, because they are very stable. Indeed, the spontaneous decay of free protons has never been observed, strengthening my string theory concept of their composition.

Therefore, we should picture neutron decay in this way:

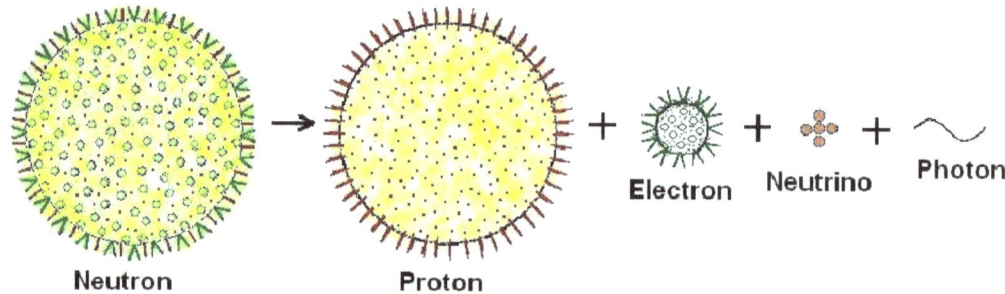

Neutron   Proton   Electron   Neutrino   Photon

Some readers might suggest that the surfaces of the neutron, proton, electron, and neutrino do not exist as I have described them—that these descriptions are just products of my imagination. However, the fact is that longitudinal and transverse waves comprised of free strings *do* exist, and they are distinguished *geometrically*. Longitudinal strings are shaped like straight lines, where the points of the strings vibrate in the direction of travel. In vibrating, they lengthen and shorten repeatedly, like springs. Again, this is one-dimensional motion, as only the one-dimensional world would exist for them. For a transverse wave, the displacement of string points is perpendicular to the direction of the propagation of the wave, so they vibrate around or even circle the trajectory of a traveling wave. These facts alone lead to the acceptance of straight-line, looped and vortical strings.

Also, forces relating to longitudinal motion (vibration) and to transverse motion do exist. The first is the gravitational force I've described above. The second is the electromagnetic force. It is well known that electromagnetic waves are traveling transverse waves. Since a circle must be two-dimensional (it cannot be one-dimensional), circled strings must provide either a magnetic or an electric force. Hence, straight-line strings existing in a one-dimensional world cannot carry electromagnetic forces.

As circles are more likely to rotate, looped strings should produce magnetism. Therefore, vibrating and spinning strings—those moving in a spiral motion (vortical strings) produce electrical forces.

When electrons are forced into protons to create neutrons (as in neutron stars), the surface strings of the protons must lie on the surfaces of the created neutrons along with those of the electrons to cancel out the electrical charges. Therefore, the neutron must have at least two kinds of string shapes on its surface.

Bosons

Given what we know of the Standard Model and String theory, bosons should consist of strings which do not create matter, and that therefore do not participate in creating mass. But they may interfere with matter for a short time, and thus exert some force on that matter. This is why we call them carriers of forces.

Free strings exist outside of ordinary matter. Originally, theorists believed them to propagate through space at the speed of light, **c**. But why not consider the possibility of strings scattered throughout space that do *not* propagate at **c**—indeed, strings at rest, with zero speed? Can we detect such strings in our environment? In fact, we can and do; we can detect their presence everywhere, even in a vacuum. For example, we can easily detect magnetic field lines, and there must be strings present to create these lines. We can't see them directly, but we can notice their effects—and we can even find ways to view them indirectly. For instance, when we place fine iron filings in a magnetic field, they line up to form visible lines that correspond to the magnetic field lines. Something must be behind this effect.

Some physicists suggest that photons create these lines. But photons cannot be responsible, because they're propagating at the speed of light in straight-line trajectories. Light bends very little over long distances, and then only in the presence of very massive objects; a small bar magnet or micromagnet won't create such curved force-lines.

That is why photons could never serve as an explanation for these continuously bending magnetic field lines. Besides, when we look at a line marked out by iron fillings, there is no movement to them. This suggests that the particles creating magnetic field lines do not possess linear momentum. Hence, in nature there should be single strings that are *not* moving at **c**—that, in fact, are not moving at all. Electric fields also display field lines, so these lines should also consist of free strings at rest.

Single strings may form many different shapes, since neither end of such a string is fixed. The basic two-dimensional shape of such a string is such that when one node is in the middle, then the antinodes are at the ends.

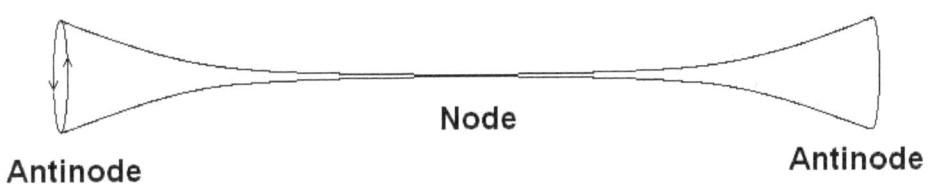

Node

Antinode   **Antinode**

If a single string spins along its length, then both ends can create conical vortices, circular in cross-section. In the case of electricity, the string's mouth is attracted to straight-line strings; i.e., a negative charge wants to be coupled with a positive charge. This attractive force may cause a change in direction. When a conical string is connected to a positive charge in matter, its mouth becomes fixed on that charge. This can cause the displacement of its node from the middle of the string to the opposite end. In this way, the other end of the string becomes a positive charge, and is ready to attract the next string's mouth. The next string behaves like the first, and in this manner a chain is created with a free positive charge at its end.

Similarly, if a negative charge is exposed, strings close to the negative charge adjust themselves to it and therefore form another chain, the end of which is negative.

When a chain issued from a positive charge meets a chain issued from a negative charge, the free ends join, closing the line and creating a electric field line between the positive and the negative charges.

Multiple parallel lines can be created the same way, emitted by the original charge and moving proportionally in all directions. If they are not interconnected, then their free ends signalize the need to be joined to an opposite charge. We may feel their presence and note that an electric charge is nearby. When the chains are closed, they create the electric field between two charges. If the positive charge or negative charge, or both, are *not* fixed, then strings creating the electric field produce a force, which attracts the positive charge to the negative charge and vice versa. These strings, then, are bosons of the electric force.

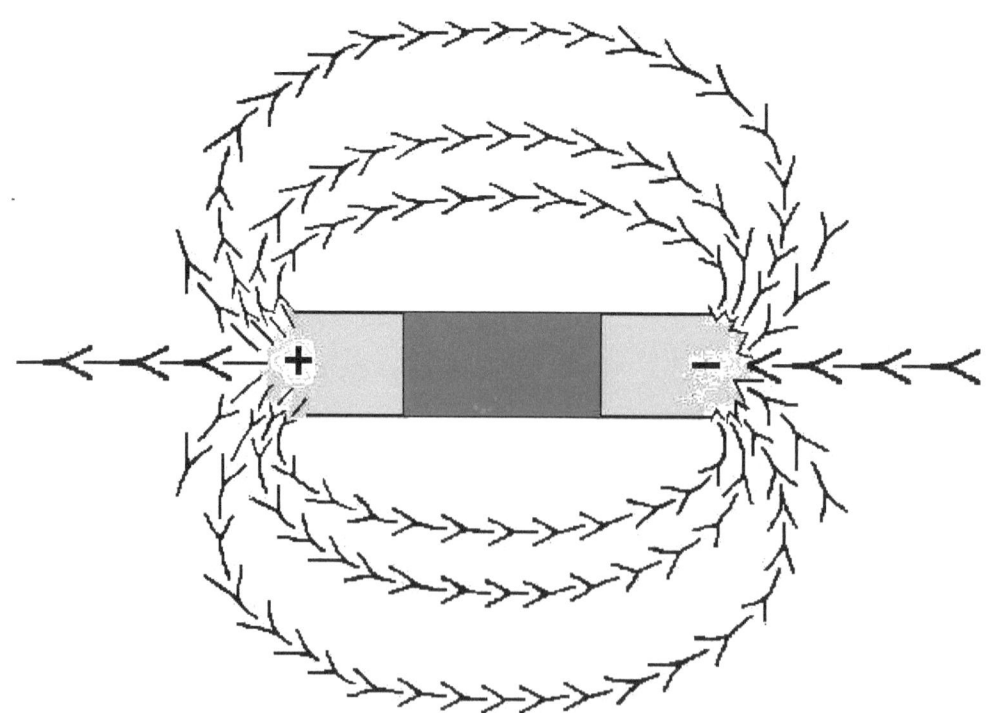

Magnetic field lines should form looped strings, since magnetic charges attract those of a different spin orientation. In the case of a single string, we know that the basic shape occurs when the node is in the middle and antinodes are at its ends. The free ends are not fixed and are spinning in circles, forming conical mouths at each end. These parallel circles are not a fit for magnetic field lines, since both have the same spin orientations. If they were, then a single string would be enough to form an elementary magnetic dipole.

Rather, I feel a single string adjusts itself to a magnetic pole so that it starts to rotate in the opposite direction relative to the direction of that magnetic pole. Another string, spinning in reverse, will adjust to our first string, and so on. In this manner, a chain forms until it meets the opposite pole or another end of a chain issuing from the opposite magnetic pole. This creates magnetic field lines. A single string becomes a magnetic monopole, and that is why a single string could also be called the **magneton**. These strings are attracted to magnetic poles; a "north" pole is attracted to a "south" pole, and vice versa. If magnetic poles are not fixed, then the strings creating the magnetic field produce a force that attracts the "north" pole to the "south" pole and vice versa. Therefore, magnetons are the carriers or bosons of the magnetic force.

It's clear from the foregoing discussion that at least two kinds of strings *cannot* move at the speed of **c** in free space; in fact, they move only to adjust themselves to nearby strings. One type would occur when strings rotate along their lengths and so create two antinodes, and the second occurs when a whole string joins ends to form a loop. Still, I don't think we need two kinds of strings here to explain magnetic poles and electrical charges. Single strings do not have any fixed ends and so are "live"; and these aspects should allow them to adjust themselves to any needed shapes. Therefore, one boson may carry the electric force *or* the magnetic force. This type of boson can occur anywhere in the cosmos in the form of scattered strings, prepared to serve for applying local strong forces.

Nature produces single strings which propagate in space by the speed **c**, which gives them linear momentum. Therefore, they can carry forces between distant objects and so attract or repel them. We should be able to deduce the forces they carry, which are necessary for the Universe to exist.

We're dealing with traveling "live" strings. The fact that these strings are live means they each possess a quantum of energy. Therefore, when matter absorbs them, these quanta contribute to the total energy of that concentration of matter. Straight-line strings contribute to the mass. Looped and funnel-shaped strings contribute to electromagnetic force, and usually to electrons, since electrons may absorb transverse (electromagnetic) waves.

These strings travel at the speed of light. Any object having speed also has momentum, and so it is with these strings. Moving objects may collide, sometimes with serious consequences. A collision of objects with different momenta results in changes of momenta for both objects. Therefore, these strings can either speed up or slow down the objects they strike—therefore, either repelling or attracting those objects. That's why no physicist should ever attempt to limit forces to gravitational attraction as a "god" of the Universe. Limiting the Universe just to attractive forces has led theorists to introduce gravitational waves to explain why the expansion of the Universe is not homogenous.

Scientists now recognize that there also exist forces that push celestial bodies apart. This casts doubt upon the existence of gravitational waves, which fail to confirm the Big Bang theory of inflation. They either confirm the existence of multiple universes or eternal inflation (a reproducible event that would happen again and again). Some scientists claim that this is so, and so identify themselves as among the pseudophysicists—because they promulgate untestable ideas. Even if these other universes existed, they would be permanently out of reach and unobservable.

Whether an object gains speed or loses it, it must follow physical laws. Newton's Third Law states that the mutual forces of action and reaction between two bodies are equal, opposite, and collinear. Since we know that transverse waves (light) lose speed when they propagate through transparent materials, then this loss must be added to the colliding partner if the collision caused no damage (that is, if it was an elastic collision).

Therefore, I propose that physics as it currently exists does not explain why adding speed to an object argues against elementary physics. We must stop financing projects looking for mysterious "dark energy," because it just doesn't exist. The physics proves this, even as the scientific numerology lose themselves in real physics.

Photons add speed as momentum to colliding objects. *Thus, the basic boson that causes objects to push apart, expanding our universe, is the photon.* The photon is a two-dimensional string, and therefore travels as a transverse wave. Photons interfere with electrons on the surfaces of material objects. That's why when they collide with electrons, they cause changes not just in the linear momentum of an atom, but also in the angular momentum of the electron (since the electron orbits a nucleus, this change may not extend directly into the center of mass of atom). This added movement produces heat as well. I discuss all this in more detail in the book *Attraction and Repulsion in the Universe*.

Straight-line strings, however, do contribute gravitational attraction as a result of their collisions with ordinary matter, picking up more speed in the collisions. The gravitational boson is the **graviton**. Both bosons—gravitons and photons—are carriers of weak forces in nature, arising only from these collision effects.

In addition, there are bosons for the strong forces, such as the electric and magnetic forces. The strings carrying these forces are local strings at rest that allow us to detect local electric and magnetic forces. Still, these "local" strings are present everywhere in the universe, governing the interactions between electrons and protons, even in a vacuum. Hence, applying these strong forces to distant objects in the universe requires carriers of electricity and magnetism.

The elementary particles carrying electric charges are electrons and protons. Their ability to travel through space is problematic, and protons at least do not propagate through matter. This is why electric force is not registered for movements of celestial bodies.

The elementary particle for the magnetic force should be the neutrino, given that its surface consists of looped strings. Its very small volume allows it to reach the speed of light. Essentially, the neutrino is as the smallest possible ball magnet. When this "magnet" arrives at a far celestial object, the neutrino may be pulled toward this object due to existing local strings comprising the magnetic field chains there. Because magnetic fields exist everywhere where there are spinning or looped strings in the structure of matter, they apply this strong force upon neutrinos and vice versa.

The greatest magnetic fields occur within the loops of chemical bonds of atoms, or in cells of crystals, and therefore take part in strong magnetic interactions. When the neutrino arrives at a celestial body where a magnetic field exists, bosons in the magnetic chain pull the neutrino inside the loop. Since the greatest magnetic field is in the center of the loop, the neutrinos will pass through the loop without encountering any atomic particles; that is, neutrinos travel unhindered through inductive matter. Besides it, an electric field around electrons and protons forces magnetism to be perpendicular to it and this, magnets avoid electric charges. Neutrinos avoid electrons.

The result of this strong magnetic interaction is that the neutrino not only accelerates due to the interaction, but the object it interacts with is pulled toward the neutrino source. This accounts for the strong attractive pull known for "dark matter."

Conclusions

I hope this work provides readers with a new way to view nature, helping them more easily visualize this world of elementary and subatomic strings. Strings teem there in uncountable numbers, all with different densities, shapes and speeds.

Strings can be packed into the densest form in a black hole, or in thoroughly smashed matter using the technology of the particle accelerator. Strings are packed so tightly there that they are unable to replace themselves. This is why collapsed matter is unable to attract gravitons, so its response to the gravitational fields of other objects is weak. On the other hand, this form of matter emits more gravitons than it should for its mass, and therefore a black hole attracts nearby objects very well.

Strings can exist bound to a firm volume of elementary particles in ordinary matter. Strings vibrate within these volumes of matter, producing the effect we know as mass. Some strings are not restricted to any volume, and move freely through the vacuum in a straight-line motion. These traveling waves propagate at the speed of light.

Strings in both ordinary matter and traveling waves have the same property of momentum, and therefore their interactions cause changes in the momenta of both partners in the interaction. The resulting effect of a sticky collision is a change in motion, and previously, we theorized that a force caused this change. When traveling waves lose speed as they propagate through a material object, we observe that the interacting object is pushed by the traveling waves, gaining speed from the source of waves. Thus, traveling transverse waves cause the photational force to act upon objects.

When traveling waves gain speed during propagation through a material object, the object is pulled toward the source of these waves. Longitudinal waves do this, exerting a gravitational force upon objects. In the case of absorption of traveling waves or inelastic collisions, the material object gains some internal energy. We register the increased movement of atomic particles—the change in internal energy—as heat.

Particles of ordinary matter have strings of different shapes on their surfaces. They can be straight-line shapes, vortical shapes, and looped or even circular shapes. The particle that includes all these kinds of shapes on its surface is the neutron. In order not to continue in shaped tension, the strings incline toward separation, and so protons, electron, and neutrinos come into existence as a result of neutron decay. Protons have straight-line strings on their surfaces, electrons have vortex or funnel-shaped strings with conical mouths on their surfaces (resulting from one free end of each string vibrating, while the other is immobile) and neutrinos have looped strings.

Nature also allows for individual strings that are not bound into any volume of matter and lack momentum. Therefore, they are *not* traveling at the speed of light, but are rather at rest. They can adjust themselves to interact with strings of different shapes if needed. For example, if one string spins clockwise, another string will adjust to it and spin counterclockwise; and so, the strings link together to create magnetic chains. If ordinary matter has circled or vortical strings on its surface, free strings join it and so create either magnetic field lines or electrical field lines.

Thanks to these individual strings, particles of ordinary matter interact with their neighboring particles, and so create atoms with different numbers of protons and electrons. Protons lie in the nuclei of the atoms, and electrons orbit them. Electrons on the surfaces of atoms may enter into bonds with other electrons thanks to individual strings, and so chemical bonds are created. The main types of such bonds are the covalent bond, which arises due to different spins, and the ionic bond, which arises due to the interaction of positive and negative charges.

Neutrons also exist in atomic nuclei. Since neutrons possess different shapes of strings on their surfaces, they can enter into direct bonds with strings on the surfaces of protons, generating the strong nuclear force that binds atomic nuclei together.

www.ingramcontent.com/pod-product-compliance
Lightning Source LLC
Chambersburg PA
CBHW040919180526
45159CB00002BA/537